Nadia BENSAÂD REDJEL

Port of Bône (Algeria): archives and documents

Nadia BENSAÂD REDJEL

Port of Bône (Algeria): archives and documents

ScienciaScripts

Imprint

Cover image: www.ingimage.com

This book is a translation from the original published under ISBN 978-620-6-71888-8.

Publisher:
Sciencia Scripts
is a trademark of
Dodo Books Indian Ocean Ltd. and OmniScriptum S.R.L publishing group

120 High Road, East Finchley, London, N2 9ED, United Kingdom
Str. Armeneasca 28/1, office 1, Chisinau MD-2012, Republic of Moldova, Europe
Printed at: see last page
ISBN: 978-620-7-96102-3

PORT OF BONE (ALGERIA) : ARCHIVES AND DOCUMENTS

This document is a quick first-hand look at the graphic and written documents relating to the construction of the port of Bône (now Annaba) under French colonial rule. From 1846 to the 1950s, the port benefited from a variety of projects and ideas, all aimed at making the town of Bône an exemplary port-city. The aim here is to share our experience of researching and using archival documents with a public of researchers and the curious. It is more important for us to offer a sort of guide to the sources where the project documents for the port of Bône are kept.We can already point to the renewed interest in historical research in Bône and elsewhere in Algeria, which will make this document an essential support for work in historical geography, the history of urban planning, or for those focusing on the interplay of institutional players.From 1832, when Bône was captured by French military engineers, to 1846, when the first plans for a port were drawn up, the engineers were more concerned with protecting the town than opening it up to the sea. Perceived as a worrying frontier, open to invasion and foreign intrusion, the sea fronts have always suffered from the disadvantages of their frontier location.For the first two decades, the shores of Bônois remained as they were, punctuated by forts and lazarets, giving them the face of the archetypal French military and historic port town[1] .

Against this backdrop, our approach involves examining and reconstructing the plans for the port facilities built during the French colonial period, in an attempt to revisit the effects of the port's proximity to the city.

Keywords : Port - Bône - Colonial Algeria - Bridges and Highways - Military engineering

TABLE OF CONTENTS

BUILDING A PORT IN THE TOWN OF BONE (ALGERIA): A STORY OF THE CO-PRODUCTION OF HYBRID SPACE

A PORT FOR TWO TRADES

The first ideas on the construction of the port of Bône date back to 1841, and the development and improvement of structures and infrastructures actually began in 1846 and continued until the end of the 19th century. The various stages in the construction of the port and the formation of the associated urban fabric can be summed up in three stages. First came the development of the small dock and the urban redevelopment that followed. This was followed by the design and construction of the large dock, the extension to the sea and the development of the maritime quarter. The Souleyre quay, the manoeuvring area, the construction of the railway station and the commercial district[2] were the final touches to the port project for Bône. The growth of the port was linked to changes in the urban fabric. The port of Bône is special in that it represents this dual reality. It is a functional port infrastructure in its own right, and is at the same time an integral part of the city, as it backs onto it.The dual construction of the port of Bône took place over several decades. It actually began with the first ideas put forward in 1841 and continued to invest more and more land until 1932-33. After a long gestation period, Bône became a key strategic and commercial centre in the French possessions[3] .

Gradually becoming aware of its strategic coastal position, Bône had a wide range of sites in which to build its port. As it was built, the port immediately adjoined the town, after the archaic mooring found there had been regularised. Since the construction of its first quay, Bône has seen a flowering of proposals for sites to build a port worthy of French power. There were two main directions, the Seybouse to the south and the Cap de Garde to the north, both protected from the ravages of the wind. From the earliest plans onwards, two government departments were involved in the port project: the military engineering department and the road and bridges department. The former, as the authority representing colonisation from the 1830s to the end of the 19th century, and the latter through its expertise. Behind the term "area" lies that of delimitation, known to the military engineers more as a strategy for securing the area than for expansion.It is important to note that the regularity of the layout of the mooring and that of the town's drainage channels stems from the same logic. Note also the precision with which the levelling is represented. Precision is becoming a sure thing here, so should the practices of the bridges and highways department be used as criteria for distinguishing between projects? What's more, requiring a project to be complete at the time meant that estimates of the work planned had to be made available. These estimates were crucial to the process of negotiating budgets with the authorities, but were not produced until much later, in 1855.

PONTS ET CHAUSSEES, MILITARY ENGINEERING: THE DIVISION OF LABOUR

And what if, on the one hand, the status of the idea took on a collective dimension through the formulation of the military genius "officers who took part in drafting the project ...", the identification of the port by the name of its builder, the Souleyre quay for example, offers a more individualised dimension to road and bridge projects.An archaic project, like those found around the Mediterranean[4] , a temporary solution, the initial project for the port of the 1840s cannot be seen as a modern structure: embryonic and unfinished, insufficient surface area, lacking in precision. The prevalence of precision from every point of view remains the key to the success of the project. This can be seen most clearly in the texts of the joint conferences[5] , to which Figures 1, 2 and 3 are linked.

Figure 1: Initial changes to the Bône anchorage

Gréban, 1849, Overall plan of the coast showing the location of the various batteries proposed by the commission for the defence of the Algerian coast in its report of 10 April 1848.

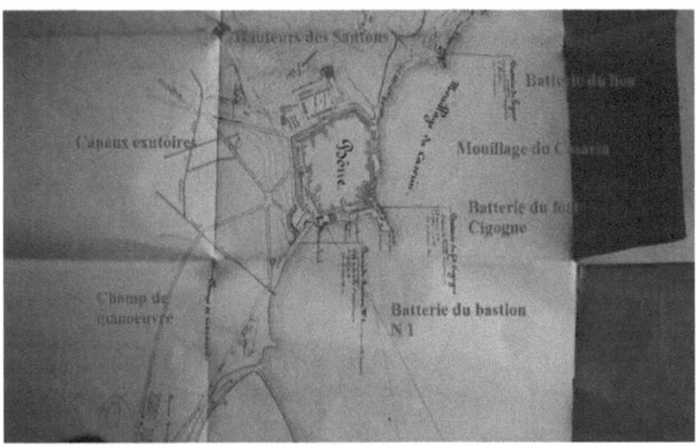

Source: 1VH 383 projects for 1849; Archives of the Service Historique de l'Armée de Terre (SHAT), Vincennes

This figure is taken from a plan of the entire Bône shoreline, showing the series of batteries proposed along this shoreline. What would become the southern edge of the port of Bône lies between the battery of Bastion No. 1 and that of Fort Stork.

Figure 2 Regularisation of the edges of the Quai Tiré du croquis de Gréban, 1878, état d'avancement du travail de bornage, 1/5000e, Bône,

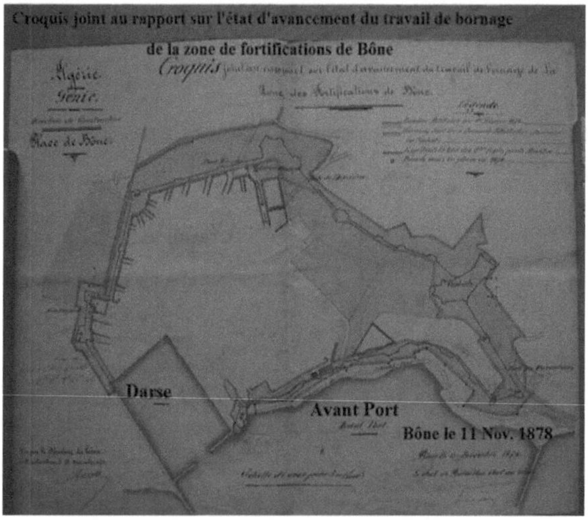

Source: 1VH 387, Archives du Shat, SHD, Vincennes, Paris.

On this plan, the layout of the port of Bône is shown in its final form, i.e. in a U-shape, and in immediate contact with the town, itself in a state of definitive demarcation: limits of fortifications, easement zones, walls and gates, the loop being closed by the port dock.

The port project only emerged in the public debate after the ministerial decision of 30 October 1852, which gave rise to a more complex and detailed project than any that had preceded it. Despite the time lag between the decision and the project, the two events are related. The written documents say so clearly[6] . The project drawn up by the bridges and roadways department in accordance with this ministerial decision was annotated by military engineering officers, who contributed to the debate out of concern for safety. Above all, projects sometimes went back and forth constantly, with the sole aim of obtaining the agreement of the Ministry of War, the legitimate authority at the time. Until then, military engineering, highways and bridges and the Ministry of War were the three players directly involved in the port's projects.

CONSTRUCTION WORK AND KEY DATES: 1855 - 1878 – 1909

Figure 3 Sketch of the port showing precisely all the depths of the small dock and the large dock; originally entitled: Port de Bône. Sketch appended to the note from the senior engineer commander dated 7 August 1854.

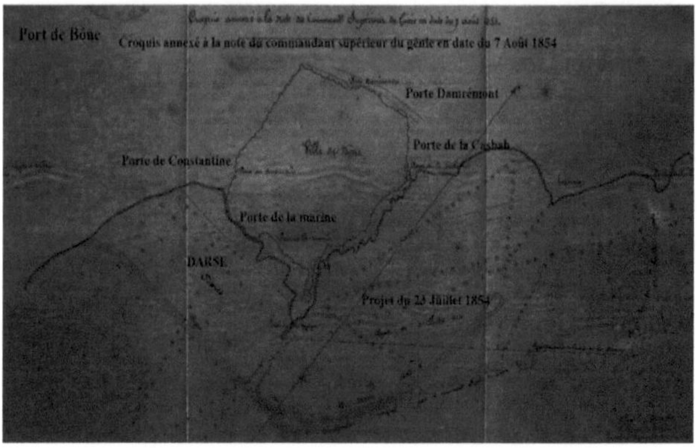

Source: 1H866; SHAT Archives; Vincennes

The figure shows the old town within its fortified limits, its four gates, and at its foot the first outline of the port. This route is made more reliable by the The sketch also shows an outline of the future outer harbour and the rue de la tranchée leading to it. The sketch also shows the outline of the future outer harbour, and that of the rue de la tranchée leading to it. The preliminary project by the chief engineer de Lannoy, validated in 1853 and completed by the final project of 1855, established the small dock covering 11 hectares and preceded by a 50-hectare outer harbour, which later became the large dock. This project ends with an urban plan showing the maritime quarter, which remained at the idea stage. The strip thus drawn is neither urban nor harbour, as it is both at the same time. It is oriented west-east and crosses the edge of the city from end to end. What kind of port space could be created in such conditions? Does this right-of-way, whether port or urban, have all the necessities of a port 'area', given that it acts more as a border? "But the orientation of the channel, with an opening of 450m facing north-east, made access difficult if not impossible [...] the Oued Boudjimah also poured its alluvial deposits into the dock, which silted up and threatened to become unusable and foul-smelling"[7] .

Work on the small dock, completed in 1870, could only be completed by diverting the Boudjimah into the Seybouse. The idea of a rear port in the Seybouse wadi was thus abandoned, and although history would later acknowledge that it was of particular interest, there was no further talk of it for decades. The years 1885 saw general improvements. The diversion of the Boudjimah, which was closely linked to the functionality of the port, also entered the cycle of projects to modernise Bône. From this series of ideas came the idea of creating a railway station, which would not be completed until 1875[8] , but before that, the priority was to extend the railway network, of which Bône was the head.Of the more or less strong relationship between all these situations, the diversion of the Boudjimah, the construction of the station and that of the port, only the port succeeded in attracting the attention of the bridges and roads department. The project drawn up by their departments did not diverge greatly from the initial idea presented above (Figure 1). Also, compliance with the codes of representation is not dissimilar to that of The same is true of the military engineering services, where the colour, legend, titles and drawings in general indicate the same thing. A distinctive feature, however, is the mastery of levelling by the civil engineering corps. Thus, the depth of the points sounded on the seabed, presented in decimetres, will remain the preferred domain of the bridges and roads as much as the drawing of meridians and parallels (Figure 4). A reminder of the contribution of this plan was signed by the Minister of Public Works, whose support was decisive for the corps.

Figure 4 drawn up in accordance with the ministerial decision of 30 October 1852 project maintained by Ponts & Chaussées 1854-55

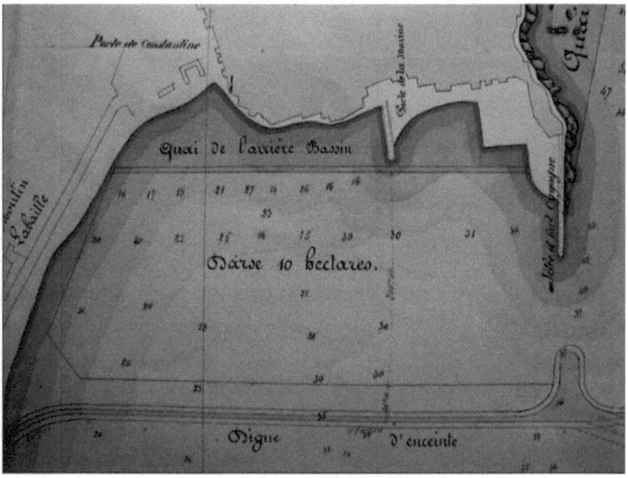

Source 1H866 ; Archives du SHAT, Vincennes ; Evolution of the Bône port project

The reliability of the port project drawn up by the Ponts et Chaussées is due to its accuracy at all points, especially with regard to water depths. In addition, the figure shows the Bône gates, the Labaille mill, the layout of the quay in the inner basin, the 10-hectare dock, and lastly, the surrounding breakwater.

MOULINIER DISCOURSES OF AUTHORITY

A simple examination of the rules of composition of this plan could raise questions about the project's compliance with the "rules of composition" practised at the time by this same body of bridges and highways, as set out in the atlas consulted for this purpose[9] .

The atlas presents an interesting number of Mediterranean port projects (Algiers, Oran, Bône, Saïd, Alexandria, Marseille, etc.), which makes it possible to compare them, even quickly. Some common features emerge (quays, docks, jetties, etc.),...) more or less stabilised depending on the projects or ports under construction, and which refer more to plans than to manuals of construction methods or specialist practices. The question is to know how much this port, which is above all a technical construction or which must ensure certain technical provisions, can also be an "aesthetic" construction.Cataloguing their work, in an atlas for example, was undoubtedly a turning point for the Ponts et Chaussées, marking the territory of knowledge that was becoming theirs, and which did not need to be too reminiscent of the technical nature of their practices[10] . Although the Ponts et Chaussées themselves did not claim to refer to these rules of the art by means of "rules of composition", we feel that it is supported by their design of the port of Bône. Does it find support in the fact that the corps belonged to the public works department, which was responsible for major state works and whose compatibility with the civil buildings department was not too much in question[11] .

The projects for the port of Bône required the engineers to give some thought to the way in which a "composition" could be articulated with the need to obtain an "area" specific to this infrastructure. There is nothing very special about this configuration, which in the long term will be reduced to a line as regular as possible and as northerly as possible, except that it has been developed over a long period of time, it has undergone various configurations, made up of progressive arrangements. The progressive advances made on the sea through the construction of quays provided an opportunity for compositions arranged to the point of delimiting a real port area, the one known since the 1870s.

The completeness of the file, "Projet du port arrêté le 4 février 1855", combined with the accuracy of its data and the follow-up it received, because it was carried out, make it a resource document, the study of which is crucial to our knowledge of the itinerary of the port of Bône. R. Lespes, who linked the development of the port of Bône to the development of the mining resources of eastern Constantine, insisted more on the tonnages that came and went on the quays than on the way in which they were configured. Are their gradual expansion really linked to the increase in tonnage mentioned by Lespes? Or has the port had to grow on the basis of a projection into the future? Lespes' work shows us just how much the port of Bône was built for iron ore. The variations in maritime traffic bear witness to this. But is its future still dependent on it? Does its expansion reflect the expansion of mining operations or the multiplication of its roles?The same thing happened to the port as to the fortifications: the simultaneity of urban events is striking here: a new enclosure was being studied at the same time as an alignment plan was being drawn up for the "new town", which alignment plan had to deal with the successive layouts of the port quays. It was difficult for us to measure the extent of the territory defined by a constantly evolving port area, which the alignment plan incorporated or disapproved of. Figure 4 clearly illustrates what was emphasised by the senior engineer commander at Bône: "We therefore conclude that this dredging project should be rejected and that the 150,000 Fr set aside by the Minister of War for the port of Bône should be applied to the southern jetty [...] dredging will be carried out if necessary"[12] .

Based on the broad outlines of the harbour projects of the 1840s, the south jetty became the determining factor in the configuration of the harbour, which was in the process of being redesigned. As a result, the final U-shaped configuration is simply the result of three successive stages, each of which traced a separate ridge: the northern, then the eastern, and finally the southern ridge as shown above. We can wonder about the reasons for such "hesitation", if in any case, the silting up caused by the Boudjimah largely contributed to the push of the sands towards the sea (Box 5).

Box 5: Extract from the register of deliberations of the Fortifications Committee (5 February 1858);

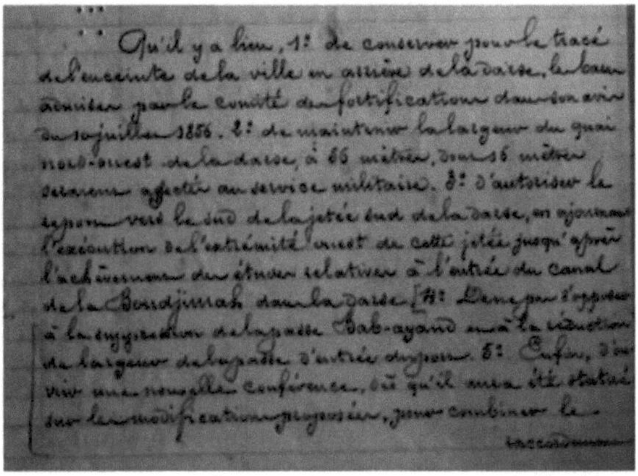
... Qu'il y a lieu, 1° de conserver pour le tracé de l'enceinte de la ville en arrière de la darse, le [illegible] arrêtée par le comité des fortifications dans son avis du 10 juillet 1856. 2° de maintenir la largeur du quai nord-ouest de la darse, à 66 mètres, dont 56 mètres seraient affectés au service militaire. 3° d'autoriser le [illegible] vers le sud de la jetée sud de la darse, en ajournant l'exécution de l'extrémité ouest de cette jetée jusqu'après l'achèvement des études relatives à l'entrée du canal de la Boudjimah dans la darse. [4° [illegible] à la suggestion de la passe Bab-ayand [illegible] à la réduction de la largeur de la passe d'entrée [illegible] 5° Enfin, d'ouvrir une nouvelle conférence, où qu'il aura été statué sur les modifications proposées, pour combiner le raccordement

Source: 1H 866; Archives du Shat; Vincennes

It is worth noting that the civil engineering corps seems to be the only body to oppose the military engineering service on issues relating to urban development. These confrontations, which the corps does not hesitate to refer to as These "protests" ended with the conferences adopting convergent opinions (Figure 7). This consensus, which was a little forced at times, is an interesting historical fact, and the archive documents reproduce it with remarkable fidelity, which is why they represent a decisive resource document for us. For example, let's mention "the minutes of a conference concerning the defensive organisation of the dock and the port", which was held at the end of the 19th century. engineering, artillery and navy on the one hand, and bridges and roads on the other[13] (Box 6).

Box 6: Ponts et Chaussées protest: Military territory or military easements versus port areas or buildings for P&C and the port;

Protestation P&C

A Bône le 3 aout 1893

Rapport du capitaine Moulinier, Chef du Génie, à Bône Au sujet de la délimitation intérieure de la zone des servitudes entre les bastions 6 et 2, à Bône ;

M. L'Ingénieur ordinaire des Ponts et Chaussées de l'arrondissement de Bône, faisant connaitre que son service désirait remplacer par une construction en maçonnerie la baraque en planche située sur le quai Nord de la darse à Bône où elle sert de bureau pour le capitaine du port, demande si l'édifice dont il s'agit se trouve dans la zone des servitudes, et, dans le cas de l'affirmative, sollicite l'autorisation d'entreprendre la construction projetée.

La baraque en question est située dans la zone unique des servitudes de la place, telle que cette zone a été délimitée suivant le procès verbal de bornage du 19 Février 1872, homologuée le 20 avril 1878.

Source: 1H 851; Shat Archives; Vincennes

The letter from Mr. L'Ingénieur des Ponts et Chaussées raises the following question of principle, on which the Chief Engineer has the honour of requesting instructions.

Figure 7 Sketch attached to the report concerning a protest by the ordinary engineer of the Ponts et Chaussées, presented in three different drawings.

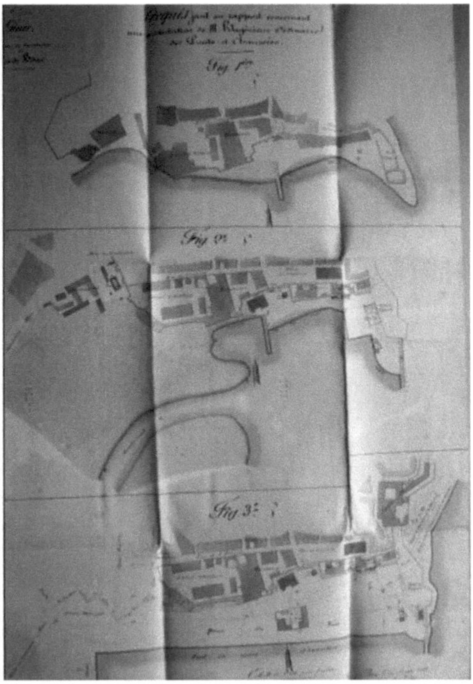

Source: 1VH 390; Shat Archives; Vincennes

The figure shows the gradual evolution of the construction of the northern quays of the port of Bône, and its advance towards the Mediterranean Sea. This represents one of the achievements of the Ponts et Chaussées and is the basis for their protest. The latter claimed the right of the Ponts et Chaussées to obtain a site for the construction of their headquarters. The only thing we generally remember about the design work of bridges and roadways is the pragmatic contribution of their works or plans. This is a very simplistic view of the quest for urban form that they can demonstrate. If we examine all the routes in their relationship to the boundaries of Bône, the purpose of all the plans drawn up by these engineers is indeed to create a material structure. The latter played a large part in the significance of the resulting urban form. It is important to understand the pioneering work of these engineers in that it broke away from the classical language of civil buildings and the pragmatic language of engineering. First of all, we had to understand what the protest was about. But this could only be done by reading the many layouts, which at first glance appear complex, but which are indicative of a serious desire to create a whole that is both urban and port-related. The resilient dimension lies in the fact that the bridges and roads department wanted to make their occupation of the port area definitive, with the solid construction of a building to house their administration (Figure 8). In their view, this was a tribute to the skills they had deployed to ensure that the port of Bône was built.

Figure 8 Ponts et chaussées building, subject of the protest;

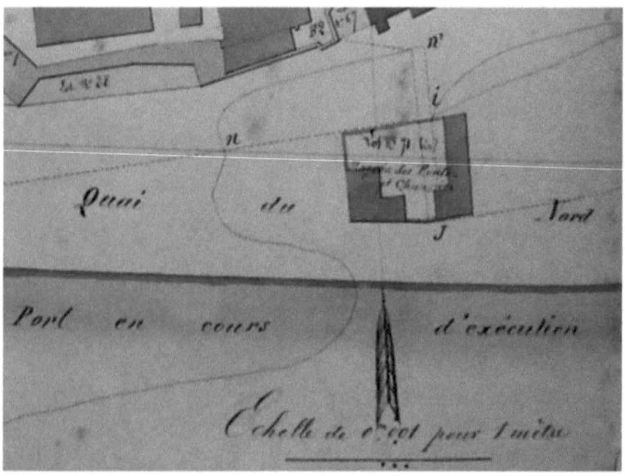

Source: 1VH 390; Archives du Shat; Vincennes

In red, in the centre of the figure, is the temporary building of the Ponts et Chaussées department, the subject of their dispute with the military engineers. The Ponts et Chaussées department claimed its right to erect a permanent building in place of the temporary one, relying on its contribution and expertise in the construction of the northern quays of the Bône harbour dock.

THE PORT, ITS PLACES AND OBJECTS OF DISCORD IN THE LIGHT OF THE PORT'S EXPANSIONIST TURN

Did the port stimulate the city or was it stimulated by the city? Also, at a time of growth in shipping movements, and this is what the work of R. Lespes points out, the port needed to expand and did Bône recompose itself on the basis of this fact? We will probably have to re-situate the debates in relation to the city-port relationship. Urban literature today is abundant and produces an enormous number of positions on the issue. In this respect, each discipline has its own entry point, its own position and its own approach. its design. Broadly speaking, there are two meanings, depending on whether the authors are dealing with pragmatic planning approaches or whether they are taking a more theoretical approach to analysing urban-port issues as an object of research[14] .

Enlargement

As it expanded, the port of Bône became detached from the town centre by building basins and quays that moved away from it. In fact, it moved away from the town centre as a result of the first boundary crossing in the history of Bône (Figure 9). No representation other than that made by the bridges and roads department in 1886 illustrates this crossing better. In this way, a new port structure was created from scratch: by overtopping the sea through a system of infilling, by the construction of new quays, and the subdivision of the areas thus gained, as well as by the opening of a dock, now known as the grande darse. This whole complex forms part of the first distinct peripheries of Bône (Figure 10). This redefinition of the city as a port, contemporary with the major growth movements, had a major impact on the direction taken by the city. Growth towards the north began to be considered in spite of the topographical obstacles, the embankments and shores, and in spite of the defensive obstacles, the boundaries and fortifications. Here's how the growth of the port and that of the town can be seen as two linked events. The need to build the new port was obvious. But then came the question of where to locate the infrastructure. A total of two sites were considered in the many proposals to house the expansion, both of them are adjacent to the port or small dock. These two opposing directions are an essential part of the port debate.The first ideas of the bridges and roadways were concentrated on the banks of the Seybouse, it seems by effects of return of the Roman history of Bône. But more concretely, river transport was not the only reason. The availability of space freed up by the filling in of the Boudjimah course, the proximity of the small dock and the opening up of the roads, before that of the railways, must have played a part in the choice of this type of site. The site is flat, but requires expensive drainage work.

Halting the expansion of the port: the Souleyre moment and the plurality of players

However, the site on which the expansion project has landed is the outer harbour, which is an indirect continuation of the harbour on the north side and is situated below the slope of the old town. It is sheltered, but rugged, and requires an advance into the sea. The prevalence of the second solution makes the new port autonomous, with a platform, huge quays and a distinction from the fabric of the town that is emphasised by the embankment. In terms of urban planning, a distinction needs to be made between the project selected, where contact with the city is certainly problematic, and this sort of opening up of growth towards sites in the north that have not yet been developed.

Figure 9 Enlargement of the port to the north, taken from the Atlas des ports de Méditerranée designed by the Corps des ponts et chaussées ;

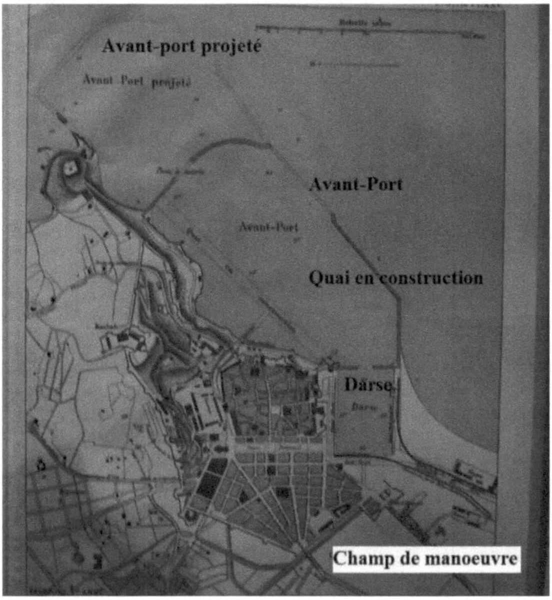

Source: ENPC, Atlas de projets de ports de la méditerranée, BNF, cartes & plans, Richelieu, Paris

The figure shows the immediate contact between the old town, its ancient fabric and maze of streets, and the port, its components and infrastructure. Linked to this ensemble is the outer harbour under construction, its quays and the rue de la tranchée. This new infrastructure backs onto the slopes of the Caroubiers, which are not yet urbanised. The city and port as a whole, conceived in a less integrated way than the current authorities would have liked, provides a new balance presented in an overall design that redefines the possibilities of the city as a whole. The project, approved by "ministerial decision of 7 October 1885", involved the creation of a new 47-hectare outer harbour to the north of the old one, enclosed on the one hand by the extension of the latter's southern jetty, and on the other by a breakwater rooted to the west at the Pointe du Lion, forming an overall convex curve towards the open sea and breaking 200m from its origin (Figure 11). The old outer harbour became the large dock communicating with the new port via a channel only 70m long. Vast open spaces forming an entire "maritime quarter" were laid out on the north side. This ensured easy access to the port and the tranquillity of the water. From this series of sub-projects, ordered within the major project of the new port, the new structure was born in a spirit completely independent of the old port: the dimensions of all the sub-spaces tend to grow, as was the case for the spaces of the basins and quays, the whole now structured into specialised functional units. Physically detached from the old port, the new port is also detached in terms of the spirit that shaped it. At the outset, no reference was made to the port's integration into the city: "this concerns a port project which does not necessarily have to be associated with an urban planning project"[15] , unlike the first dock, which was included in the various plans for alignments in Bône, this one would not be linked to the city until much later, by a simple trench street, the motivation for which has been explicit since the initial ideas behind it.

Figure **10** Project chosen for the new port of Bône;

Source: 1H 856; Archives du Shat; Vincennes

The red lines indicate the housing estates and therefore the future functional town, which will be made up of land reclaimed through the progress of the seafront, as indicated by the planning procedures. These "exceptional polygons" were eventually conceded to the administrations of the This is the "port of Bône". The speed with which this progress has been made must be emphasised, as much as the city and port encourage each other. Pragmatism, a lack of resources and the topographical difficulties of connecting the Grande Darse and its quays to the city overshadowed the geometric qualities of the route forced into regularisation (Figure 10), and it would take time to achieve a synthesis of the city and port composition. A gradual abandonment of the aesthetics of the plans in favour of a utilitarian design based on a repetitive grid would suddenly eclipse the efforts made by the bridges and roads department in their previous project, as illustrated in Figure 58, which shows the complexity of the regulatory routes. In order to create a street that, paradoxically, would become the most structuring feature of the port complex thus created, they had to set themselves apart by proposing a succession of particularly disparate architectural styles. The link between the isolated maritime quarter and the city was gradually developed via Rue de la tranchée (Figure 12). At its meeting on 20 February 1883, the Fortifications Committee issued its opinion on the project to enlarge the port in accordance with the ministerial dispatch of 25 January 1883. The joint works conferences (engineering and roads and bridges) of 16 November 1882 and 9 March 1880, as well as the drawings and plans attached to them, formed the basis for this opinion. The latter was requested by the Minister of War, who returned all the documents: the project to expand the port of Bône, supplemented by the organisation of a deliberation by the town council "accepting the charges imposed on the town" and modified by a new variant proposed by the chief engineer of bridges and roads. The complete file includes A conference report dated 16 October 1882, with drawings attached; a report by the head of engineering on 26 October 1882, followed by the opinions of the director and the general, the senior commander of engineering; a supplementary conference report at 2ème level dated 3 December of the same year, with 2 drawings attached, and finally a decision by the town council.It is probably unnecessary to stress that it is the long duration of the debates that emerges from this chronological sequence. This was caused by the struggle between the engineering and road departments, and even if a consensus eventually emerged in the form of resolutions, these remained difficult to achieve. Even though the organisation of joint conferences by the engineering department was a sign of a certain spirit of conciliation.All the town-planning measures taken by the engineers subsequently had repercussions on the other departments: the division between the port and the town was once again an opportunity to take steps to impose the power of the engineers on the town:"The Committee, after having heard the reading of the report which specifies and having deliberated, agrees with the rapporteur that the military service can accept without inconvenience the trench leading to the quay in the extension of the Boulevard du Nord, with its new width of 20 metres, as well as the side ramp to this trench shown on the sketch attached to the 2nd degree conference of 3 and 12 December 1882, on condition that all the costs of building the engineering workshops, the canteen and the enclosing wall of the mixed quarters, remain the responsibility of the civil service; and consequently that the War Department should also agree to the exchange of the land referred to in the minutes. and the deliberation of the Bône town council dated 18 December 1882 in the courtyards of the Damrémont and d'Orléans districts"[16] .

Figure 11 Treatment and regulation of the new dock;

Source: 1H 868; Archives du Shat; Vincennes

One of the skills of the Ponts et Chaussées (civil engineering) department lies in its precise knowledge of the depths of the water in the small and large docks. And this is what Figure 11 shows: written in red, "the regularisation of the bottom at coast - 08 m"; the points concerning "general dredging". Further up is the quay and "earthwork under construction". Of particular note is the large area of land reclaimed from the sea."They wanted to think big, and events justified their forecasts and answered the concerns of those who were worried about the disproportion between the work and the port's actual traffic, and who foresaw a large expanse of water with quays but no ships. In 1885, the same year that work began on the To carry out the programme adopted, Ph Thomas discovered the lime phosphate deposits in the Tébessa region"[17] .

OPENING OF RUE DE LA TRANCHEE

The opening of Rue de la tranchée is in fact nothing more than a vain attempt to link the port to the city, and the image of this unity conveyed by the plan is, in our view, illusory, because in reality the street does lead to the port, but with this action the latter appears as a separate entity (Figure 12). Moreover, there is a remarkable discrepancy between the importance of this issue and the way in which the projects have responded to it, summarily resolved by a fragment of street within the major port project.

Figure 12 Opening of rue de la tranchée

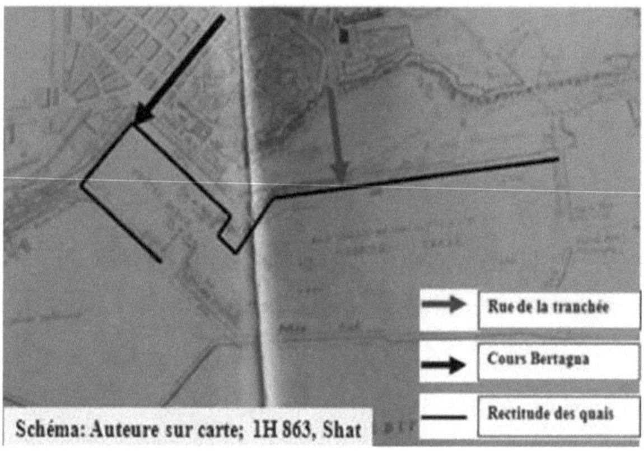

Both the Cours Bertagna (orientation in black), and the opening of the rue de la tranchée (in grey), the orientation of the spaces in the town of Bône is the same. Note the prevalence of the orientation of the structure of Bône towards the sea, making it a resolutely port city. "The outer harbour offered the largest surface area, an anchorage from 8m to 14m deep. The work carried out from 1885 to 1894 and interrupted for lack of funds from 1894 to 1899, was resumed at that date and on the eve of the war, the programme could be considered to have been carried out. The port was completely safe and in excellent condition; from 1906 to 1919, commercial operations were only interrupted by the surf for three full days. By 1901, the quays were 1150m wide in the Grande Darse and 600m wide in the Petite Darse. But until recently, the equipment remained rudimentary. While the quayside lanes were sufficient for the traffic, as were the platforms, the same could not be said of the only shed belonging to the Société des Phosphates de Constantine, on the Petite Darse quay, and the sheds belonging to the shipping companies"[18] .

Figure 13 View of rue de la Tranchée ;

Source: Bibliothèque Nationale de France (BNF), Département cartes et plans, estampes, Richelieu, Paris.

There is no need to dwell on the practicalities of laying out and rebuilding the fortifications, which were so well mastered by the bridges and roads department. The rampart road, which is extended by the trench road (Figure 13), is laid down as a stable condition, in the face of temporary roads opened for the needs of the worksite, designed to bring stones and sand to the site. Another condition set by the engineers was that the re-establishment had to be discussed at a joint conference (engineering and bridges). Who would authorise the construction of either type of road, temporary or permanent?"The condition is to re-establish the continuity of the rampart street by a carriageway, or at least one that can be used by horses, and to beat the approaches to the outside"[19] .

All this is simply the consequence of the project to demolish the santon hill. This era was rightly distinguished by the profusion of projects of this type, so complex because they answered a multitude of questions in a single gesture (Box 14).

Box 14: Joint conference report;

En résumé, le comité est d'avis :
Que le ministre de la guerre peut adhérer au projet d'agrandissement du port de Bône qui a d'abord fait l'objet des conférences mixtes du 16 octobre 1882, puis de la conférence au 2ème degré des 3 – 12 décembre suivant, provoquée par l'ingénieur en chef des Ponts et Chaussées pour l'examen d'une variante sous les réserves suivantes :
1° - Tous les travaux à exécuter au bâtiment militaire qui borde la tranchée et les rampes à construire seront faites par le service du génie aux frais de l'administration civile.
2° - Il en sera de même pour les travaux de reconstruction des fortifications ; Le service des Ponts et Chaussées restant toutefois chargé de l'exécution des fermetures des portes en avant du fort Cigogne et aux débouchés de la tranchée, dont l'établissement est intimement lié à ces travaux.
3° - Tous les projets de détail de ces ouvrages seront rédigés par le service du génie et examinés dans des conférences mixtes, où l'on fixera également le montant de la somme à fournir par le service civil pour pourvoir à leur exécution. Pour copie conforme

Le colonel du Génie	**Le Général de Division**
Secrétaire	**Président**
Signé Henry	**Signé Brossonnet**

Source: 1H 866; Shat Archives; Vincennes

In the box, the three conditions laid down by the engineers so that the Minister of War could give his approval and the work could go ahead, show the different ways in which this department could compromise with the town in the first instance as far as expenditure was concerned, and with the highways and bridges department as far as responsibility for execution was concerned; the decisive detail was well kept in the hands of the engineers.While this statement may reflect a far-reaching attempt to reach a consensus, we will find later on that the division of responsibilities is far less fair than the box shows. For in reality, the enormous quantity of excavated material resulting from the removal of the santons' ridge was the sole responsibility of the highways and bridges department, a decision that would lead to the more sensitive issue of moving the fortifications, an issue to which the engineers were reluctant to yield. "On the other hand, it will be necessary to allow the removal of the part of the fortifications that is in the way"[20] .

But the new route had to be studied immediately at the conference This was to be done "in accordance with the instructions given by the engineering rapporteur, after the road and bridges department had marked out the exact limit of the excavation to be made, so that the construction of the new enclosure could be completed even before the excavation reached the existing enclosure"[21] .

On this question of the division of tasks, our investigations cannot lead us to answer at this stage either by the conflict of competences or by the search for compromise, as both states of affairs had to come into play in turn depending on the circumstances. Generally speaking, however, it was the engineering services that monopolised the decision-making aspects the most, thanks to the control they exercised over the land from the early years of colonisation. Neither fortifications, as a security issue, nor competence, as the bridges demonstrated their mastery, enabled them to achieve this.

EXPANDING THE PORT AND SUBDIVIDING THE TOWN IN A SINGLE MOVE

The question of the delimitation of the quays of the port of Bône was the subject of lengthy debates in a series of conferences. This may seem confusing, because in any case, the quays were laid out, built and therefore delimited from the outset. In this case of clear physical boundaries, what was it that needed to be demarcated?Thus, at its meeting on 13th April 1872, the coastal defence committee issued an opinion on the minutes of the conference held on 13th August 1869, concerning the demarcation of the quays in the port of Bône, a question which gave rise to an opinion from the fortifications committee, dated 6th July 1870. Between the start of the debates in 1869, followed by the various opinions issued by the fortifications departments in 1870, and 1872, the date on which the coastal defence committee was held, speculation increased on the land that seemed to be available, and the long cycle of this delimitation made it all the more coveted (Figure 15).

Figure 15 Drawing attached to the report by the railway engineers, Bône - Ouenza Port - Railway - Allotment of quays; Maritime districts and urban allotments, Existing old town and future old town (in alignment plans).

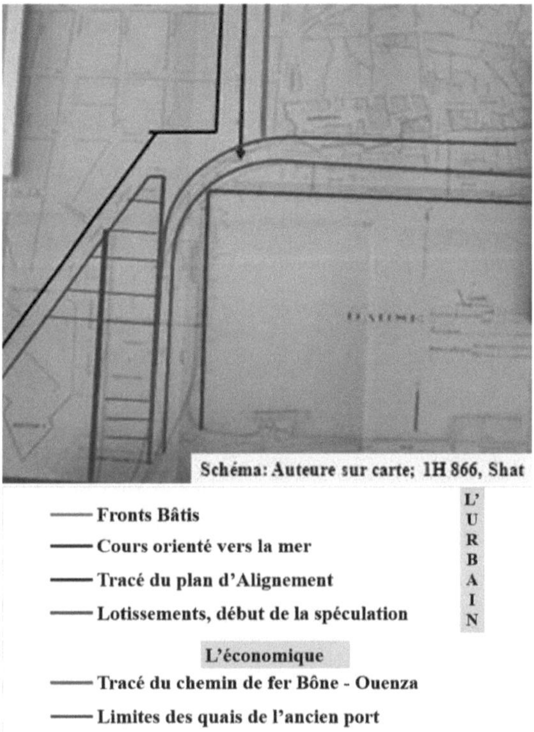

Source: 1H 866; Shat archives; Vincennes

On 22 May 1872, the commission's conclusions were as follows: The question of the location to be reserved for the station of the railway from Constantine to Bône via Guelma "had to be postponed until the layout of the fortification which was to link bastion no. 7 to the defensive wall of the southern jetty had been decided at a conference"[22] . The pre-eminence of the layout of the fortifications and their actual construction is very clear here. The layout of the station would then depend on this, as would the demarcation of the platforms, which was probably the next logical step. However, this sequence of events did not have the effect observed by the mixed conferences without the port project: the quays were set at 35 metres in length, and the quays and embankment were now the only point of contact between the water and the town, which is a testament to the result expected by this series of conferences. It was also necessary to decommission the part of the old fortifications between the rue Napoléon and the stork battery tunnel, and immediately demolish the Porte de la Marine. The successive compositional structures of the north quay show how unstable a project remains during its realisation and through its confrontation with the realities on the ground. Negotiations led to the provisional acceptance of certain alignments for the arsenal and materials handling establishments, which were themselves provisional.But what seems more significant is the location of the temporary building, which seems to have gone through various administrations before finally becoming the office of the port director: "Finally, to allocate to the military navy the land that this department now occupies on a temporary basis at the east end of the north quay, reserving for the war department the right to erect on this land the buildings that will link the stork battery to the defensive wall of the east pier"[23] . His final assignment was to the port director, who was none other than a civil engineer. Once again, this body seems to have been the only one to confront the effects of engineering power on the city. Recourse to the reservation procedure was also often used: "questions relating to the relocation of these establishments (the arsenal and the handling) and the final alignment to be adopted for this part of the north quay will be the subject of subsequent studies and conferences"[24] . To conclude, the questions submitted and discussed at successive conferences

BOUDJIMAH DIVERSION

For its urgency and for the need to build the station that depends on it, or rather on the land reclaimed by this diversion. All the members of the joint committee agreed on both issues. "M.M. the engineers should therefore be invited to complete the project being studied for the diversion of the Boudjimah and to examine what measures should be taken to apply to this work the special resources coming from the sale of the developed land, either to the new company for the station, or to the town of Bône, or to the chamber of commerce or to private individuals"[25] . The phrase "Land sold to" refers to a variety of themes, all undoubtedly linked to urbanisation. But more specifically, the ideas already expressed about development as the spearhead of colonisation are particularly relevant to the city and its inevitable expansion. Putting considerable areas of land gained through the diversion of the Boudjimah on the market and making it available to the town or even to private individuals gave a new twist to privatisation, the main adversary of the engineers, who were the sole holders of the land until then. In the final analysis, and to consider only the physical aspect of this large-scale operation, the development by bidding up the value of the banks of the roads created by the same procedure is a method that has often been tried out in the metropolis[26] and whose direct consequence was speculation (Figure 66). Here is the significance of this. Of course, it is vital that the old bed of the Boudjimah is filled in. "We also agree with the provisions proposed by the Generals commanding the engineers and artillery concerning the construction of a parapet 4 metres wide at the top and 2.5 metres high on the left bank of the rectified bed, using the excavated material"[27] .

Creation of a second dock

As the project presented several years ago by Mr Celler the engineer has to be considered as definitively abandoned, the administration will study, when the time comes, a new project taking into account the indications provided by the mixed commission concerning the location of the new dock, but the question must be reserved as it seems established that the current dock (1875) is sufficient for the requirements of trade once it has been deepened to its full extent.

Building the new station

The committees felt that it was absolutely essential to open up to commerce and industry all the state-owned land to the west of the dock (now known as the Champs de Mars), and to maintain free communication between the quay and the Porte d'Hippone, which gave access to the Guelma road. As a result, the new station had to be built entirely outside the fortifications, on state-owned land stretching as far as the new bed of the Boudjimah. In any case, this solution no longer poses any problems, as a result of the arrangements agreed between those in charge (commanding generals of the engineers) and those of the artillery. The latter had already had to work together to complete the enclosure of the fortifications of the square from bastion 7 to the west quay of the dock, measures that would only result in the civil services having to carry out an inexpensive job. The solution was to give the land over to commerce and industry, even if it meant placing the station outside the walls at the risk of, to avoid the station attracting attacks from the natives. The layout indicated for the new station is a reproduction of the one studied by the railway control engineers and the engineer from the Mokta company; this sketch plan, drawn up in support of the conference, was handed over directly by the Prefect after he had consulted him (Figure 16). The new station would have an area of around 7 hectares; it is likely that this area could be reduced to 5 hectares for the time being, according to the forecasts of the department's general council; large proportions are planned for all the railway facilities, so that all traffic requirements can be met once commercial transactions have developed as expected. The tracks assigned to the goods service inside the station would be connected directly and without turntables, which is very important, with the west and south platforms of the dock; there would even be a site near the corner of these two platforms for the establishment of sidings, which would greatly facilitate the movement of goods. The space reserved in the interests of defence, between the station and the fortifications, would justify the creation of a new square at this point, which would provide convenient access for passengers and contribute to the embellishment of the town. The mayor of Bône insisted with the engineers that the station street, which would run parallel to the platforms, should not be obstructed by any overhanging buildings; the proposal for a street 15m wide would not be exaggerated given the likely movement of cars bound for the station. In addition, all the streets perpendicular to the platforms should be 12 m wide, as it is important to open up large outlets in an area where most of the local commerce will soon be concentrated. It is important to emphasise the concerted but mechanical construction of the streets and squares in the commercial district. "We have shown on the attached plan the route of the major public roads that could be created between the new station and the Guelma road. These are simple indications that give an idea of the use that could be made of the state-owned land to be granted and in no way prejudge the question of subdividing this land, a question that should be studied as soon as possible by the departments concerned"[28] .

Relief from military easements.

On 22 February 1875, the Minister of War authorised this rebate under conditions which appear to be compatible with the requirements of civilian services.

Land subdivision.

As shown on the plan, the development to be adopted for the land to the west of the port is mainly for commercial purposes and will be put up for sale immediately. It is interesting to note the extent to which the port is part of the town and part of the development. Lots 31, 32, 33 and 34, which run along the north quay of the dock, will be leased by mutual agreement between the bridges, engineering, artillery and general inspectorate of civil buildings in Algeria. Algiers, 16 March 1875, "We have the honour of informing the Fortifications Committee that, by today's decision, the Minister has approved these matters as far as the War Department is concerned. Opinion of the Inspector General of Civil Works, Port of Bône, access roads, subdivision of land and construction of a station, in agreement with the joint conference "[29] .

However, it is essential to remember the contribution of the Inspector General of Civil Works, which was more related to urban works (housing estates, roads, squares) than to those of the Port itself.

Figure 16 Drawing attached to the Chief Engineer's report dated 23 October 1876; the drawing shows the major new harbour project, including a back basin to collect water from the outflow canals, the railway station, urban housing estates and the town moving inland.

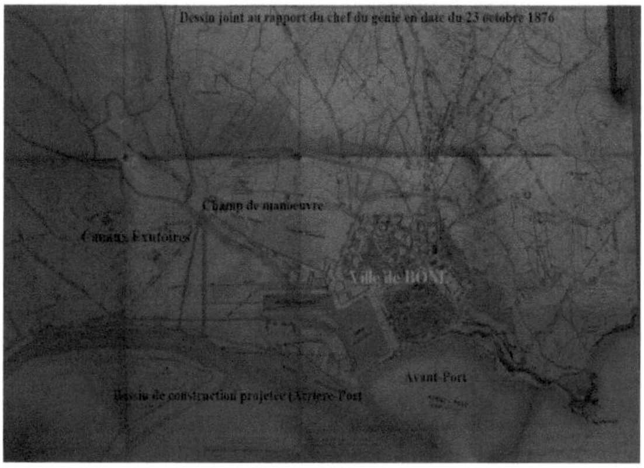

Source: 1H 866; Shat archives; Vincennes

The city-port of Bône and its surrounding area are shown in its entirety in this diagram. Above all, it shows the ambition to integrate the urban and peri-urban areas and to include countless projects, including those for the inner harbours within the city. valley. Bône is a city of water, with marshes and canals in the Seybouse valley, harbours and the sea on the fringes. Figure 17 summarises some extremely complex facts. In fact, just by reading it and respecting a certain gradation in the scale of the elements to be read and transcribed, the map shows above all how the "exceptional polygons" fit into a port project, but also into the urban whole. We understand the extent to which this area has continued to be the subject of special treatment and even covetousness, so much so that it alone brings together the potential of various branches of the city's economy and development. But more than the physical spatial fact, the micro-territories that Moulinier takes up are precisely delimited, because each one will know the assignment that suits the logic of the body in charge of appropriating it. The drawing, which in itself argues the division into micro territories, is illustrative:

is the "regulating" route the cause or the effect? As it happens, what in our explanatory approach corresponds to the search for and subsequent highlighting of territories (urban, port, micro ...), is nothing other than what the genius refers to as "exceptional polygons". These too are the subject of lengthy and complex debates between the players involved. Even the engineering corps seemed to be divided between the artillery, the navy, which had nothing to do with the port undertaking, and the fortifications; let's not forget the predominant role played by the highways and bridges department in the entire implementation of the port (design, construction, operation). Respectively, each of these bodies intended to defend its hold on a given territory both through the force of the law, and the interventions of the Minister of War, who sometimes played the role of mediator to ease tensions, but also through the expertise in the area that characterised it during this long period of implementation. The stretch of water was inevitably the responsibility of the navy, which sometimes had to intervene in matters of coastal security; the part coloured yellow on the map in figure 18 was the subject of long disputes between the bridges and the artillery; as for the polygon coloured purple, it remained the most coveted micro-territory for various reasons: security, for the fortifications and the navy, operation for the port, development for the town and by the latency of a hitherto discreet body "the civil buildings" surprisingly supported by the bridges and roadways which have led the fight on this plot to free it from military servitude, more so than other bodies[30] .

Figure 17 Sketch of the boundary of the easement zone; in violet, the exceptional polygon free of all easements, definitively ceded to the management of the "Port of Bône" in 1893.

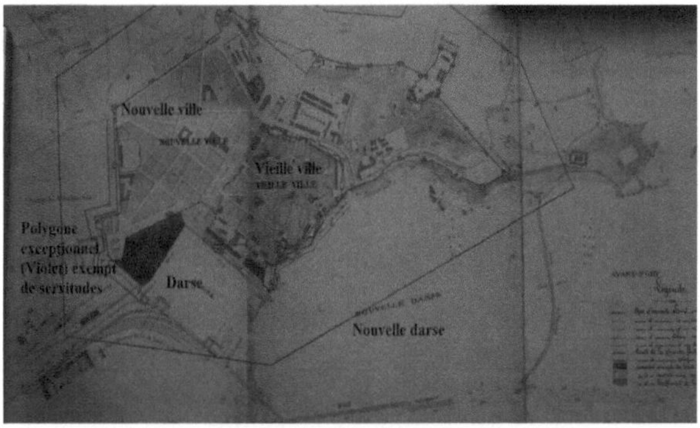

Source: 1H 862; 1893

One of the exceptional polygons used to build the port is shown in purple. An exceptional polygon is nothing other than an approach that is itself exceptional in order for the military engineers to cede part of their land territory to the benefit of other government bodies (City, Roads and Bridges, Port, Navy, Artillery, etc.).

CONCLUSION

Like all ports located in urban areas, the port of Bône is part of both a port and an urban issue. What's more, here, Bône and its port identify with each other. Its itinerary represents one of the themes of the city's evolution, from several points of view.The studies carried out by the bridges and roads engineers, which began in 1849, are certainly based on arguments with technical connotations and also show their weight in terms of decision-making, but they have informed us as to what a "project" can be for this powerful body. In relation to the need for knowledge which guides our investigations, the moments of implementation of the project and its concrete realisation or execution, are distant, and thus bring us back to the questions of the weight of ideas on the actual creation of spaces. From then on, the project only takes effect on the basis of its prerogatives and procedures, which should be seen as the sole driving force behind the creation of the port.Even the promises of prosperity for Bône through the expansion of its port did not go so far as to authorise the bridges corps to derogate from its rules of composition, which had to be respected. The art of composition, which we usually understand as the interplay of geometric configurations and conformance to the classical rules practised at the time, has encouraged us, along with the project practices of the bridges and roads department, to broaden this vision to the dynamics of adapting geometric layouts to the technical and functional imperatives of a port. This did not prevent the port of Bône from also being an "aesthetic" construction, but not as a priority. From this point of view, the culture of atlases should be seen both as a reference for the skills of bridges and, above all, as a catalogue revealing an order of possibilities. In this case too, the long duration of the debates had an effect. The struggle between the engineering and road and bridge departments not only established the idea that consensus should prevail, but also, in crucial situations, asserted the primacy of the latter over the former, whereas the opposite was often accepted. Bône's relationship with its port played a large part in structuring the whole. But when it comes to the division between the port and the town, once again an opportunity to make arrangements, this port system provides Bône with a territory with a plural profile. We see it in terms of all the opportunities it offers, both economic and urban. It may seem paradoxical to see it as an urban planning opportunity, because the new facilities (the maritime quarter to the north and the Souleyre terreplein to the south) are known solely as port structures. They were an opportunity because of the urban expansion they provoked, which can only be linked to them: diversion of the Boudjimah, establishment of a second dock and a railway station, relief from military servitudes, establishment of land allotments.In our opinion, the fact that these systems are port-based and specialised in the strict sense of the term should not obscure the dynamic they provide, even if within them they have been fragmented into economic units with performance requirements.

DOCUMENTS

Appendices (I) - Transcripts of joint conferences

Report by Battalion Chief Grimaud, Chief of Engineers, on the request to decommission certain parts of the enclosure and its extension (Requested by the General Inspector of Engineering in his inspection order in 1884)

Note that there is a direct link between the decommissioning of the enclosure and the expansion of the port. The Bône town council having expressed the wish that the north quay of the Bône port dock should be widened by moving the military establishments fronting on this quay and that, as a result, front 1-2 should be demolished, the Civil Governor General of Algeria ordered on 10 July 1872 that this question should be examined at statutory conferences held between the various departments concerned and in which the representatives of the town council and the chamber of commerce would be heard. These conferences took place on 30 July, 14 August and 4 December, and on 17 May 1873, the Minister of War announced that the town was not in a position to make any explicit proposals, and that there was no reason to follow up these requests.On 18 November 1879, the town council once again passed a resolution calling for the relocation of the military buildings located on front line 1-2. A dispatch dated the following 17 February from Mr. Le Général Commandant Supérieur du Génie ordered a new conference on the subject, which did not take place until 29 April 1881. At this conference, the Head of the Engineers reminded the 1er deputy, representing the municipality, that the Minister, in his dispatch of 17 May 1873, had made it known that he had not followed up on the 1872 conferences, because the town had not been able to designate the land on which the new military establishments that were to replace the old ones could be built, nor to indicate when it could pay the construction costs.As no proposals were made at this meeting, it was necessary to meet again on 28 December

1881. At this conference, sites were designated for rebuilding the handling and artillery arsenal, but none were designated for the subdivision of the Engineering Headquarters.At this last conference, the municipality extended its demands, proposing to demolish the enclosure from the Karézas gate to the dock and to replace it with another enclosure that would encompass the manoeuvring area, the fodder yard and all the land to the left of the Constantine road as far as the diversion of the Boudjimah, where the stations of the Bône-Guelma and Mokta-el-Hadid companies are located. In a dispatch dated 10 May 1882, the Minister, after examining the proposals made at this conference, announced that he was purely and simply maintaining his decision of 17 May 1873. As a result of these various conferences, the town requested the demolition of front 1-2 and all the military establishments attached to it, before the demolition of the part of the enclosure between the Karezas gate and the dock, and undertook to move the latter forward, so as to extend the town as far as the Hyppone bridge. It also undertook to bear all the costs of rebuilding the military buildings and the new town wall, on condition that all the military land that would become available in this way would be handed over to it; it would replace the current manoeuvring area with another manoeuvring area near the square, for which it would bear the purchase costs.Front 1-2, which runs from the arsenal to the artillery to Fort Stork to Bastion 2, still carries a servitude on the North quay, albeit tempered by the creation of the exceptional polygon on this quay. It is certain that the easements on frontage 1-2 are not motivated by any defence considerations, as the North quay, and consequently the town, are sufficiently protected by the stork fort, which closes off the square and under which there is a defensive passage for communication, and the width of the dock, which prevents any surprises from this side.The forthcoming decommissioning of this front was provided for in the ministerial decision of 29 July 1876, approving the committee's opinion of 27 June 1876, and the Superior General Commander of the Engineers in Algeria, in rejecting a request to build, admitted in his dispatch no. 61 of 11 January 1879 that Front 1-2 would soon be decommissioned.

Under these conditions, the engineering department is asking for this downgrading, not only in the interests of the civilians who will benefit from it, but above all in its own interests, to get rid of a cumbersome surveillance system that generates a lot of paperwork.The intention is to modify the enclosure from the Karézas gate, encompassing the manoeuvring area and the forage park, and to extend the enclosure along the left bank of the Boudjimah diversion. The reasons given to justify this extension are :

1° the lack of building land in the current site. You only have to look around the city to see that there are still plenty of building plots, single-storey and ground-floor houses that will be built in the near future.

This will provide housing for future population increases.

2° facilitating communications between the town and the stations, protecting these stations against enemy incursions, and eliminating easements on this side of the stations.

Communications will not be any easier; the stations are reached via the Hyppone road and several exits on the south quay, as the removal of the enclosure means that no additional exits can be built. Today, the stations are cut off by the Boudjimah river to the south and by the fodder park enclosure to the south-west, an enclosure that is defensive or easy to make so; finally, the manoeuvring area forms an open esplanade that is easy to monitor, thanks to the line of houses beyond the area, on the Hyppone road. This terrain, beaten by the fodder park and the 11-10 font, as well as by this line of houses, is not easily accessible to the enemy who might want to surprise the stations.

This means that the stations end up protected from any attack, the easements are not necessary and it would be possible to satisfy civil interests by removing them and creating an exceptional polygon in front of the front extending from the dock to the Porte d'Hyppone.

3° cleaning up the manoeuvring area;

The faubourg Saint-Anne, which is already covered with buildings, is no healthier; the fumes from the surrounding plains and the drainage sluices will always exist, and the agglomeration of the population on low-lying land where no sewers can be built will, on the contrary, create a new source of infection. What's more, it's hard to say when these plots of land will be able to be covered with buildings, when plans to create a maritime district on the outer harbour side will result in business moving to this area.

Thus, from the point of view of civil interests, there is no advantage to be gained by extending the enclosure, and even less from the point of view of defence. The project for the new enclosure, as requested, would extend the square on the side of the heights most at risk, that of Bou Hamra and Boughantas, which would necessarily have to be occupied by permanent works.

The city will extend in all its length, no less than 3 kilometres from the new Hyppone Gate to the Caroubiers Gate; the removed enclosure is only 880 metres long, whereas the proposed enclosure would be approximately 3170 metres long, i.e. 2290 metres longer; this would give the perimeter of the square, including the Kasbah, a development of 7kilometres 800 metres. This was a very large area to be defended by the mobilisation garrison, currently set at 9,000 men, including 700 convicts, and comprising only 580 men from the active army and other corps.A city that is still sparsely populated, with a heterogeneous population that is partly Italian and not all equally safe, can only be safe if there is a modestly sized enclosure that is easy to monitor and easy to navigate.

Extending the city walls weakens the city because of the extent of the ramparts, the distances to be covered, the difficulties of surveillance and the greater ease of panic. Today, as in the past, a town needs an enclosure encompassing the core of the city and detached posts outside, occupying the important points; in this way, the most pusillanimous part of the garrison can remain within the enclosure and the good ones in the advanced posts. If everything is included in an enclosure that is too vast, the defenders will have to be more numerous and one wonders whether, in the event of war, the squares of Algeria could be provided with sufficient garrisons if all of them, as is generally requested, were to extend beyond the current limits.

If we were to remove part of the enclosure and extend it, it might be appropriate to remove it altogether, as it would no longer be of much value. This extension would be a burden for the war, with no effectiveness for defence, and as for removing it, it would always be time for the population to suffocate in this enclosure, as the saying goes, which cannot happen any time soon. Finally, without any conference or request on the subject, the town would like to see the disappearance of the sea fronts from and including Fort Stork, in addition to Front 1-2, or at least Fronts 9, from the Casbah Gate to the Bastion 31 salient.

This question was submitted to the Inspector General of Engineering in 1879 who said, in his order left in place:

"This measure would have the disadvantage of eliminating all security on the seaward side, as the two piers of the outer harbour, which leave a 400-metre gap between them and which, contrary to the original project, have received neither a defensive organisation nor a battery on their muzzle, cannot be considered as an addition to the enclosure. The Head of Engineering will endeavour, in the conferences that will be held on this subject, to give the greatest consideration to civil interests and in particular to facilitate the extension of the exceptional polygon; but he will specify the maintenance of the seafronts as one of the conditions for the support of the engineering department".

At the last conference, held on 16 October 1882, on the project to enlarge the port and create a new maritime quarter in the outer harbour, located in front of the seafronts, the civil services did not requested the demolition of these fronts. The exceptional polygon created on the land in front of the seafronts, allowing construction up to a height of 15 metres, seems likely to satisfy all needs. The question of demolishing the seafronts therefore does not seem likely to come up again for a long time.

Statement of the extension project

In spite of the limited success that the project to extend the walls of Bône may have with the military authorities, as a result of the considerations set out above from a defensive point of view, we are going to examine the various details of this project and the consequences that would result from it. The current enclosure, which needs to be moved, leaves a gap of 170 metres between the dock and bastion 7. This gap has been left despite the project adopted by the Minister on 3 August 1875, following the various projects under consideration to enlarge the dock and build the Bône-Guelma railway station inside the enclosure, on the western quay of the dock.

The new enclosure, which the municipality is asking to be built to replace the one between the dock and the Karézas gate, would start from this gate, which would be preserved, follow the left side of the Philippeville road and be attached to the enclosing wall of the fodder park, which would be preserved as the enclosure of the square, albeit with less resistance; It would then cross the Constantine road near the Hyppone bridge, where the Hyppone gate that had been removed would be rebuilt behind it, and would then follow the left bank of the Boudjimah diversion and the seafront to connect with the jetty that closes off the entrance to the dock in front of Fort Stork.

In this last part of the project, the enclosure was drawn parallel to the seashore, although the waters are rising a little each year inland, which has necessitated the abandonment of the railway battery; plans were made by the bridges and roads department to dam the right bank of the Seybouse at its mouth, continuing the damming up to the southern jetty of the outer harbour; This embankment is necessary to protect the Mokta-El-Hadid railway station, which would otherwise be flooded sooner or later; once this embankment has been completed, the enclosure could follow the project route and even be brought forward if the embankment reaches the sea; otherwise, the route would have to be kept much further back, starting from the planned battery at the mouth and the diversion of the Boudjimah, on a line parallel to the above-mentioned station. Inside this new enclosure, an internal fortification zone would be kept, which would serve as a rampart street, and the armament of the removed part would be transferred there by building 3 batteries, one of which, at the mouth of the Boudjimah, would contain the 6 pieces of 4-piece field artillery currently placed between the darse and bastion 7, intended to beat the plain on the right bank of the Seybouse, the course of the Seybouse and the Calle bridge over this river, as well as the roads leading to it. This battery would be set up as a small enclosed square, with the fodder yard forming another. A second battery would be placed near the new Porte d'Hyppone, and would continue the two parts currently at Bastion 7.

Appendices (II) - Transcripts

APPENDIX PV Bône Port Routing Conference / Defences 1858

COPY EXTRACT

FROM THE REGISTER OF MEETINGS OF THE FORTIFICATIONS COMMITTEE

Meeting of 5 February 1858

Bône

Minutes of a conference on the layout of the port of Bône and the organisation of its defences See the plan annexed to the conference minutes of 29 November 1856 The Minister of War referred back to the committee for examination the minutes of a conference held in Bône on 29 November 1856, between representatives of the artillery, engineering, navy and highways departments. This conference on the final layout of the port of Bône and the organisation of its defences had been called for by the Fortifications Committee in its opinion of 10 July 1856, as being essential for the study of the layout of the seafronts of the new walls of this town. However, in this opinion, the committee indicated to the chief engineer the bases of the project to be studied for enclosing the town on the side of the planned dock and for linking the southern dyke of the port with the new fortifications, by opening up the port in accordance with the provisions laid down in a ministerial decision of 4 June 1851. In accordance with these instructions, the Chief Engineer proposes that the edge of the north-west quay of the dock and the curtain wall behind it be laid out parallel to the alignment given by the Labaille mill, and first, from this mill, 18 metres for the width of the military street, 1 metre for the site of the defensive wall, and 40 metres for the quay of the dock, in addition, he requests : 1° that this quay remain entirely free of any permanent construction or deposits, so that nothing impedes the defensive action of the town's perimeter wall, 2° that the western end of the south pier be built, from point 8, located 35 metres from the location of the pan coupé at the bottom of the darse, until we that a special conference be held to study the connection between the southern dyke and the square's enclosure and the planned connection with the dock of the Boudjimah canal; 3° that the musoir of the southern jetty be increased to 32 metres wide to accommodate a battery and that the musoir of the Bab-Ayaud mole be laid out in such a way as to accommodate a defensive structure flanked on one side by the Lion battery and on the other by the southern jetty.The artillery commander accepted the defensive arrangements proposed by the chief of engineers; he also indicated that the battery on the Bab-Ayaud pier should be closed at the throat, equipped with seven cannons and contain accommodation for 50 to 60 men, and that the battery on the south pier could be equipped for five guns with a guardhouse and shop. The civil engineer agreed with the chief engineer's proposals, with the exception of the prohibition on building on the dock quay; he asked for complete freedom to build sheds to house goods, as well as the premises needed to set up surveillance posts. In addition, the necessary space for a port light should be provided at each muster station.The port director also requested authorisation to build health offices, boat repair and reconstruction sheds, and shops for equipment and fuel depots on the dock quays. Before closing the minutes, the members of the conference agreed that the 150-metre pass between Bab-Ayaud Point and the eastern jetty should be abolished, that the jetty should be topped by a wall at least six metres high, and that the southern jetty should be moved further south on the banks to extend the quay and make it possible to establish a beaching beach needed for repairing ships and boats, and finally, that the southern breakwater be moved further south onto the banks, to extend the quay and allow for the establishment of a beaching beach needed for ship and boat repairs. The directors of artillery and fortifications approved the proposals and conclusions of the conference minutes, in particular the removal of the Bab-Ayaud pass. The chief engineer for highways and bridges was opposed to reducing the space allocated to the dock by building the north-west quay wall 56 metres from the Labaille mill; he asked that this distance be reduced to 45 metres, including 12 metres for the rampart street, 1 metre for the fortification and 32 metres for the quay. As for the other modifications proposed to the port project accepted by the Minister of War, he felt he had to reserve his opinion until experience had been gained. provided insights that would enable the question to be resolved satisfactorily. However, he had no objection to widening the muzzles, but on condition that the resulting excess expenditure would be paid from the funds specially allocated to the engineering department.

With regard to the buildings to be erected on the quay, the chief engineer asked to be allowed to freely set up all the sheds, shops and lime kilns necessary for his department during the construction of the port, with an undertaking to empty the premises at the first requisition of the military authorities, and in any case after completion of the port. In addition, he supported the proposals made by the ordinary engineer and the port director to reserve sites for the shelters required for trade and port surveillance; however, he felt that these shelters should be made of wood, not fixed to the ground, and subject to the condition that they be demolished in the event of war. The senior engineer commander approved the layout of the north-west quay and the enclosure behind it, as proposed by the chief engineer, noting that a bank with a slope should be built inside the enclosure wall, reducing the actual width of the military street to around 12 metres. It also adopted the provisions presented for the port's defences. The removal of the pass at Bab-Ayaud point, and the postponement of work on the western end of the southern breakwater.With regard to the site required for the installation of the harbour works contractor's yards, this general officer pointed out that the question had been dealt with at a special conference held on 4 April 1857, so there was no need to raise it again here. As for the constructions required for public services, which it is proposed to establish along the dock, the senior engineering commander, while admitting their usefulness, believes that they will have to be the subject of a special decision by the minister, in accordance with the provisions of the decree of 10 August 1853 on defensive easements.

The senior artillery commander approved the location and composition of the proposed batteries, should the port project be maintained. He also felt that further conferences were necessary, either to examine the modifications to be made to the existing land batteries, or to study the artillery resources to be used to protect the port, in the event of the general modifications requested for the port.

Appendices (III) - Transcriptions - Enlargement port 1883

COPY EXTRACT

FROM THE REGISTER OF DELIBERATIONS OF THE FORTIFICATIONS COMMITTEE

Joint works

Meetingof 20 February 1883 Bône

See attached drawings and plan At the conferences of 16 8bre 1882

And of 9 March 1880

Committee opinion on the port expansion project, issued in compliance with ministerial dispatch

25 January 1883

In a dispatch dated 25 January 1883, the Minister submitted to the committee for examination a file concerning the project to enlarge the port of Bône, supplemented by the organisation of a deliberation by the town council accepting the charges imposed on the town and modified by a new variant proposed by the chief engineer of highways and bridges.

The complete file includes

A conference report dated 16 8bre 1882; and appended drawings

A report by the Chief Engineer dated 26 8bre 1882, followed by the opinions of the Director and the General, Chief Engineer.

Supplementary minutes of the $2^{\text{ème}}$ conference dated 3-12 December of the same year, 2 drawings appended.

A decision by the town council

Appendices (IV) - Cartographic sources on the port

Port of Bône [Map]

GE DD- 23 (166)

Rade du fort génois et port de Bône [Cartographic Document]

GE FF- 2738 (83)

Documents - Appendices (V) - Procedures

How should urban history be categorised, and which spaces should be covered?

The port is probably one of the areas structuring Bône.

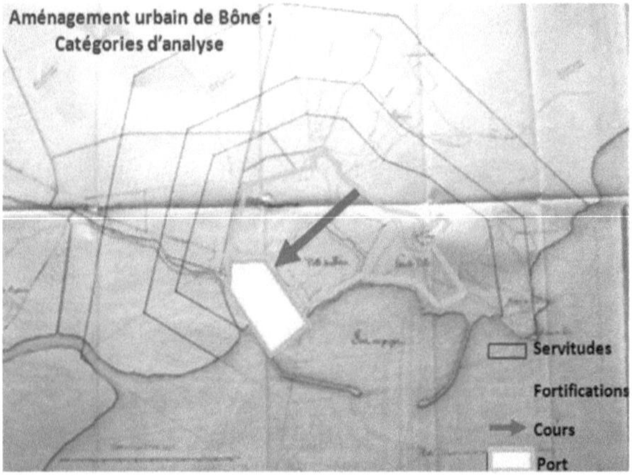

Source: author's diagram on 1878 map of Bône

Deciphering: the exceptional polygon of the small dock on the west side / Two readings (Engineering and Bridges)

Jacquier (P&C) 13 Juillet 1893 Moulinier (génie) 3 Août 1893

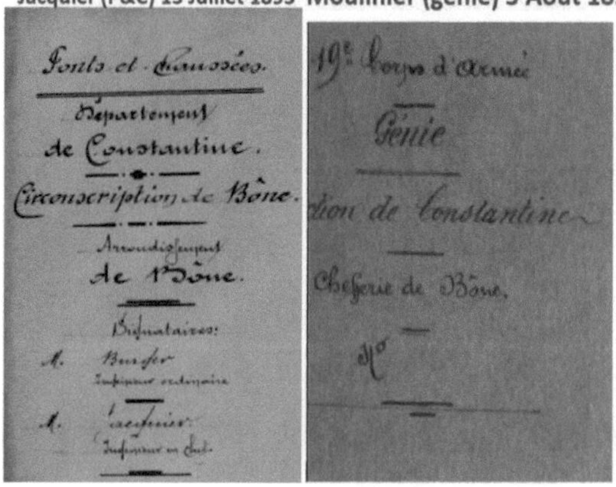

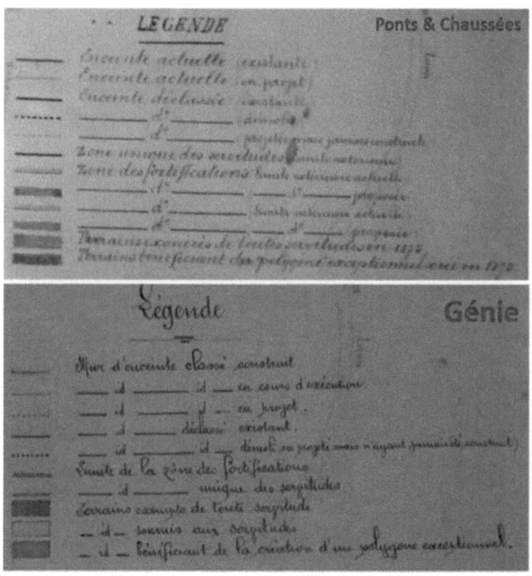

Appendices (VI)

Station and port: a functional and structural pairing

Les avants projets présentés par la compagnie concessionnaire du chemin de fer de Bône à Guelma

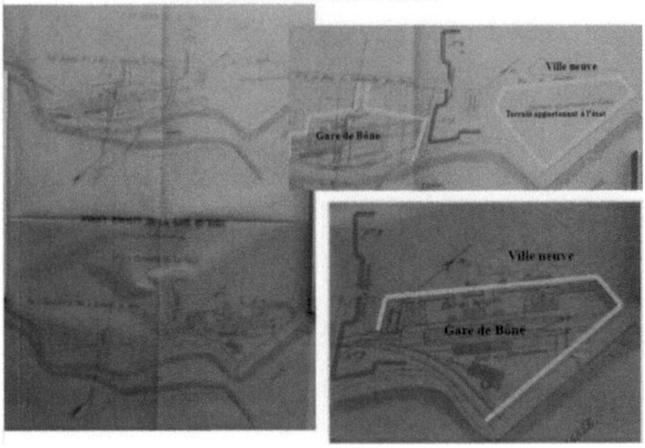

Appendices (VII)

On the "First ideas for a port" page, acquire the reflexes for using the archives, to what extent you should look: Authors, Dates, Titles, Signatures and agreements, Sources and codes; all essential information for using the archives as precisely as possible. Accuracy does not mean assertion.

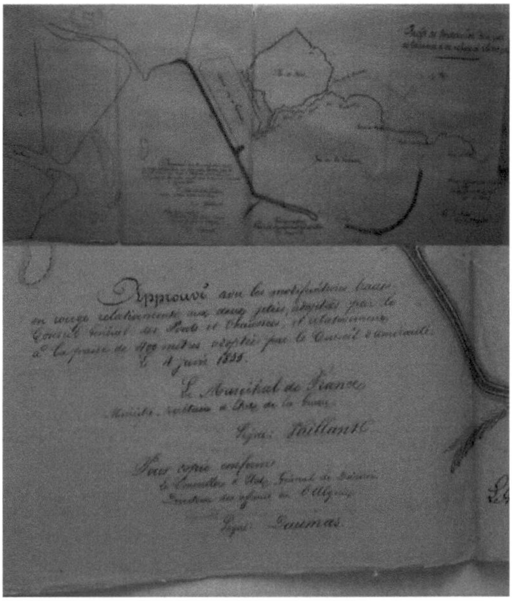

Source: 1H 866, Shat, Vincennes

Appendices (VIII) - Evolution of port projects for Bône

1850s (top left), first draft of port project; 1860s (top right), relative precision of port project;The 1870s (bottom left), when the layout and depths became more precise; the 1880s (bottom right), when the city-port project took shape.

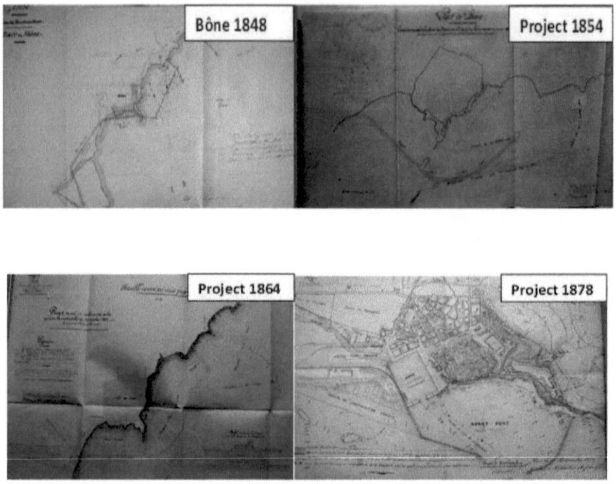

BIBLIOGRAPHY

Archives of the Service historique de la défense (SHD), Service historique de l'armée de terre (SHAT), Sub-series 1H 851; 1H 854; 1H 862; 1H 863; 1H 866.

Atlas de projets de ports de la méditerranée, Ports étrangers, 1885-1886, BNF, Département cartes & plans, Richelieu, Paris

Bensaâd Redjel, N., (2009), (Under the supervision of), Final projects in urban planning, Department of Architecture, Université Badji Mokhtar Annaba, Algeria.

Debrie, J., Eliot, E., Soppé, M., (2005), Un modèle transcalaire des nodalités et polarités portuaires, Mappemonde, N° 70, 3-2

Debrie, J., Eliot, E., Soppé, M., (2009), Des ports maritimes pour quels territoires? A geographical look at port decentralisation, Conference given at FIG. 09.Debrie, J., Eliot, E., Soppé, M., (2013), La relation ville-fleuve-port dans les projets urbains : acteurs et instruments d'une reconnexion (Strasbourg, Lyon), Cybergeo : European Journal of Geography [Online].

Gaudin, J. P., (2000), Desseins de villes : Art urbain et urbanisme, Editions L'Harmattan, 174 pages

Graff, Ph., (2000), L'exception urbaine : Nice, Ed. Parenthèses, 171 pages

Lespès, R., (1923), Le port de Bône et les mines de l'Est constantinois, Annales de Géographie, t. 32, n°180, pp. 526-541.

Masson, F., (2014), Trois revues institutionnelles : Le Journal de l'École polytechnique, les Annales des Mines, les Annales des Ponts et Chaussées, Revue de Synthèse, Vol. 135, Issue 2-3, pp 255-269

Prelorenzo, C., (2011), La ville portuaire, un nouveau regard, Rives méditerranéennes, 13-22.

Revue Chantiers Nord Africains, July 1930, January 1934, January and July 1936, January 1937, BNF, Paris, France.

MIX
Papier aus verantwortungsvollen Quellen
Paper from responsible sources
FSC® C105338

Printed by Books on Demand GmbH, Norderstedt / Germany